SPACE

COLORING BOOK

FOR KIDS

THIS BOOK
BELONGS TO

ASTRONAUT

SPACE SHIP

ALIEN SHIP

SUN

HALF MOON

FALLING ROCK

ALIEN FACE

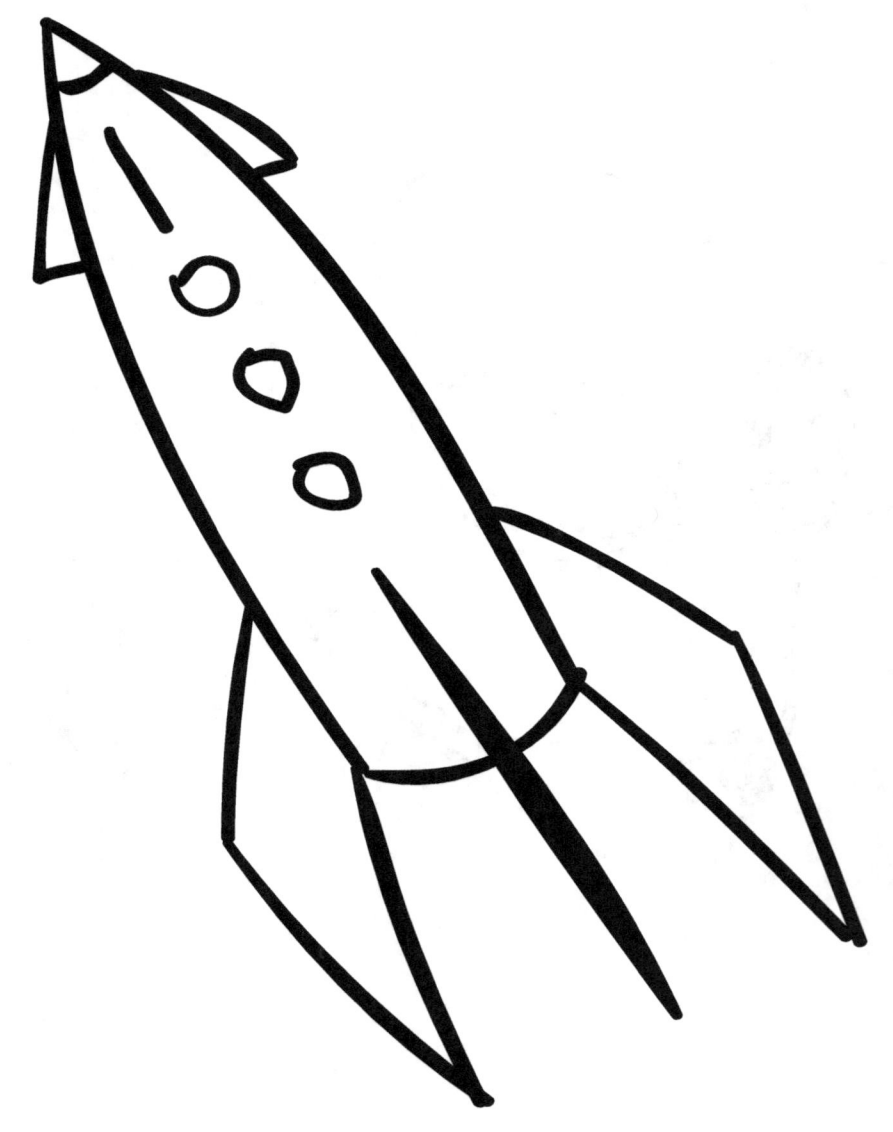

ROCKET

FALLING STAR

PLANET EARTH

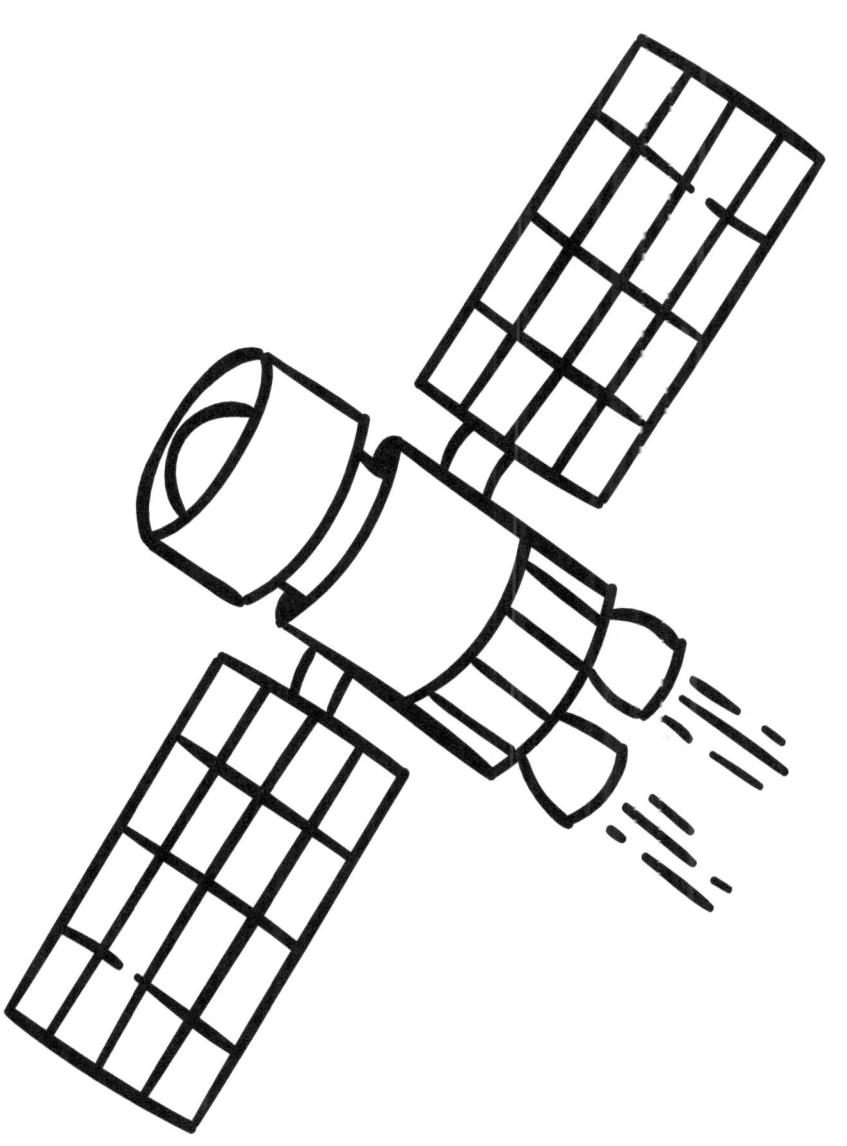

SATELLITE

SATURN

FLYING SAUCER

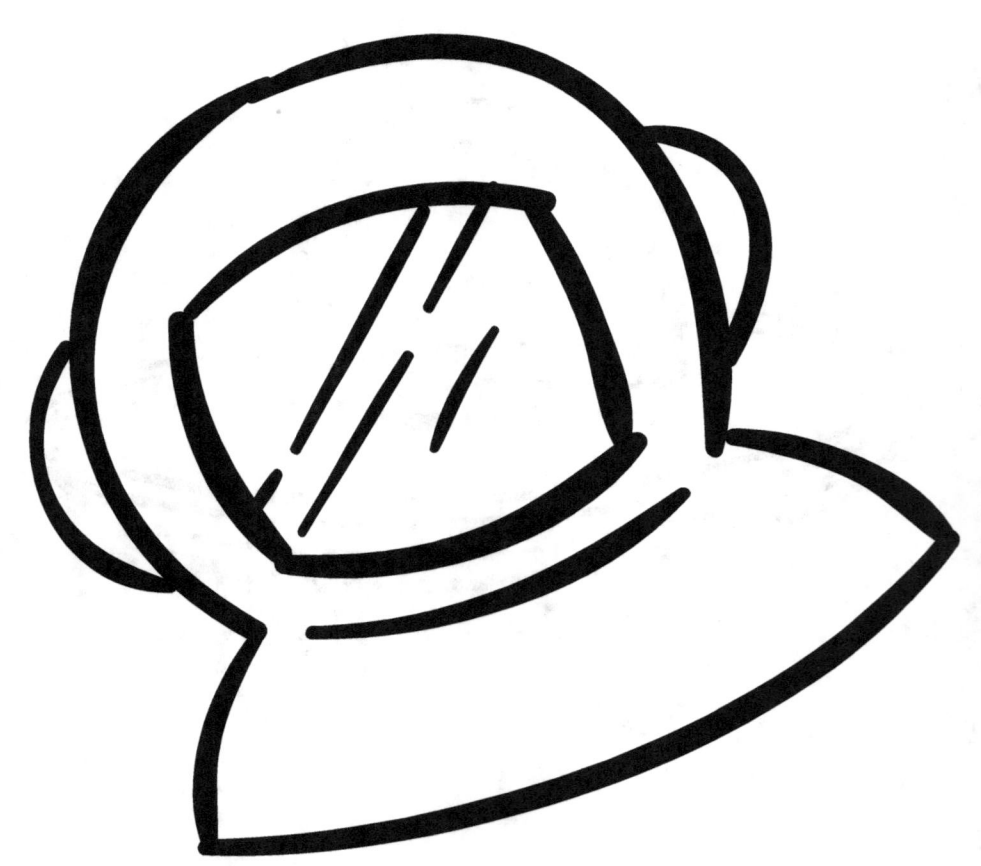

ASTRONAUT

CLOUD

FLAG ON MARS

FALLING STAR

PLANET

ATOM

FLYING SAUCER

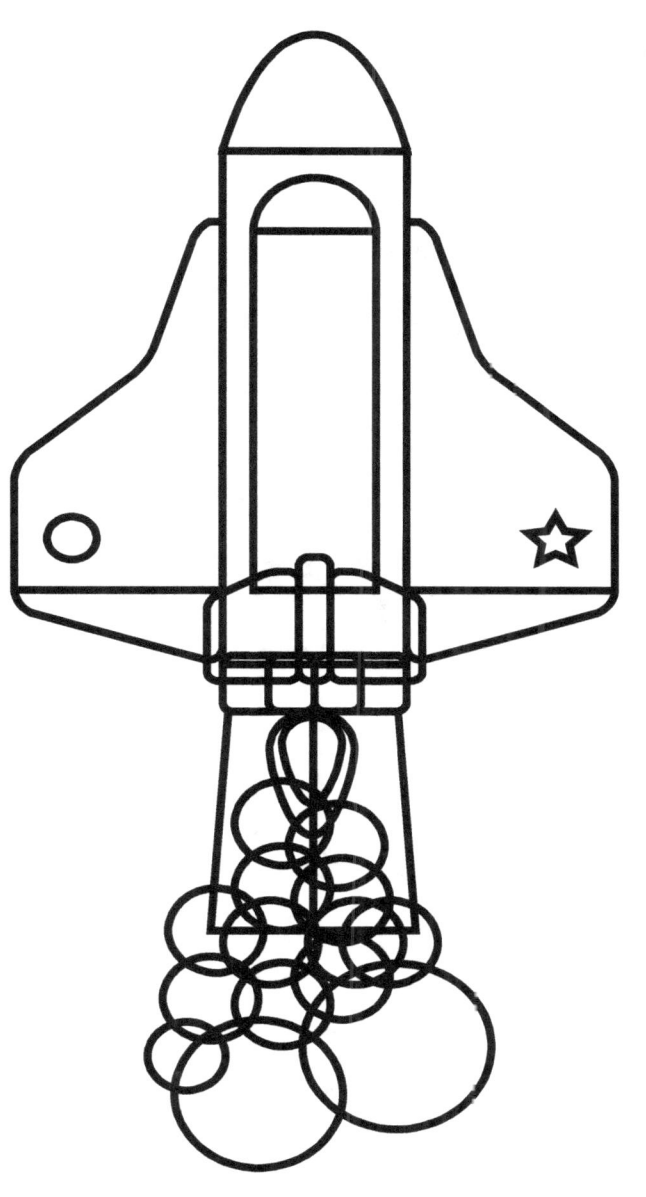

SPACE SHIP

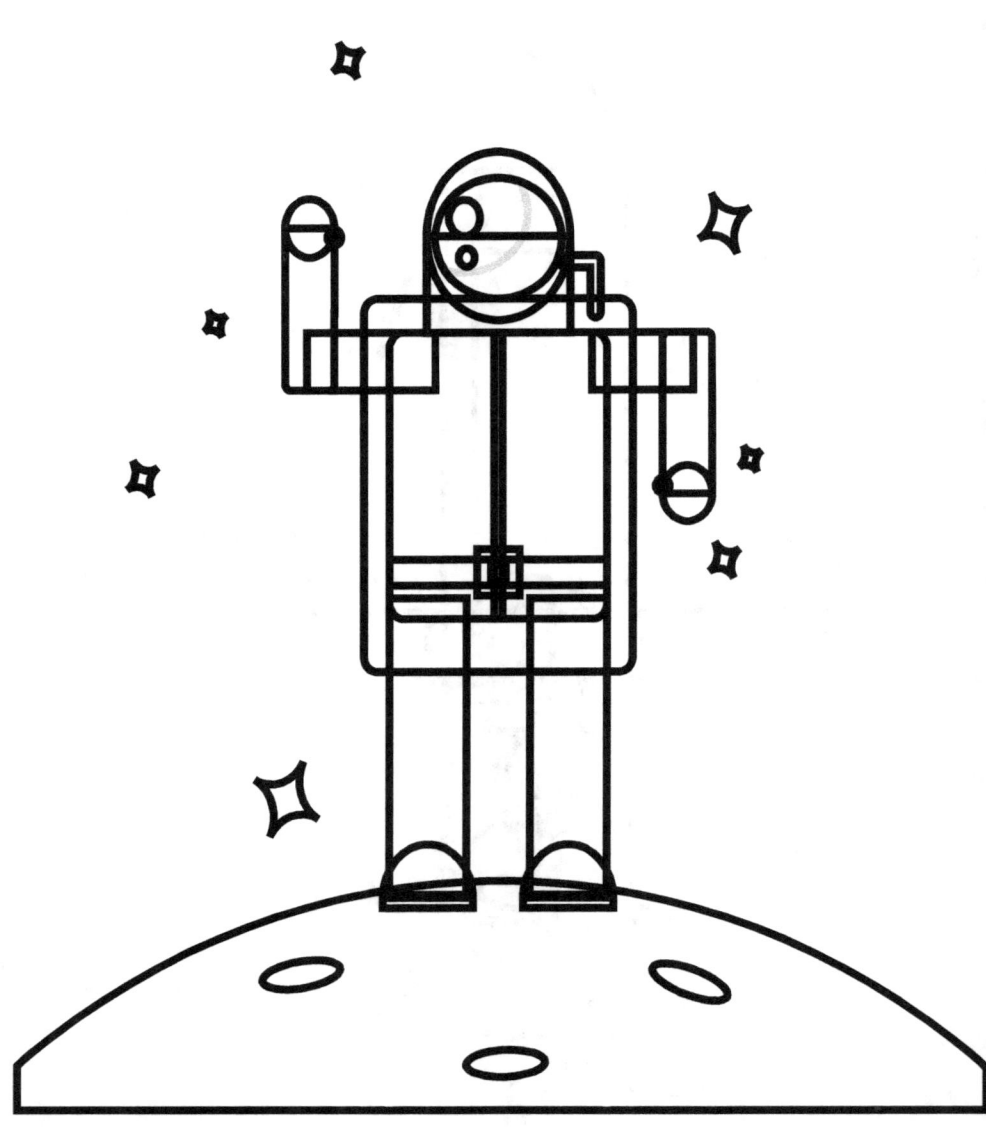

ASTRONAUT

SATELLITE ANTENA

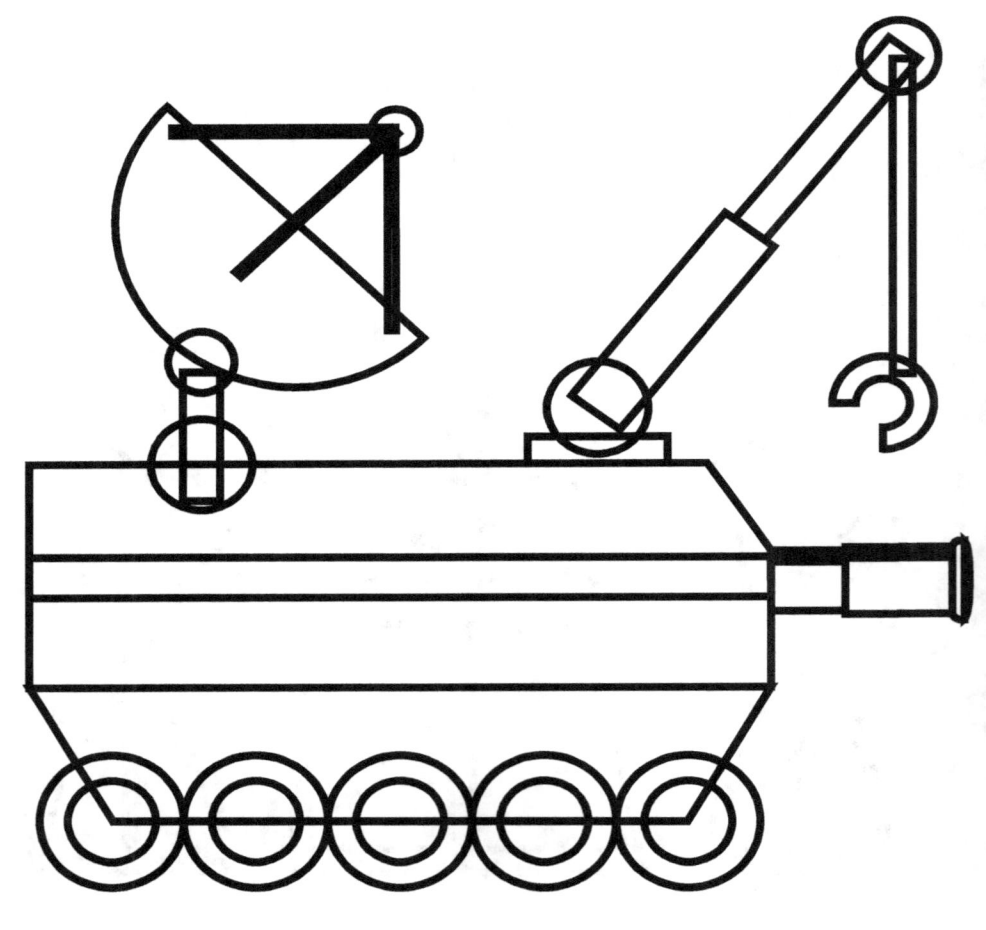

SPACE CAR